WORSE THAN

THE HALIFAX EXPLOSION

M. Pauline Murphy Sutow

Acknowledgements

The publisher wishes to express his appreciation for the generous financial support of the **Nova Scotia Department of Tourism and Culture** and the **Canada Council.**

Richard Rogers, Publisher

FOUR EAST PUBLICATIONS
P.O. Box 29
Tantallon, Nova Scotia B0J 3J0

1st edition August 1992

edited by Grayce Rogers
layout and design by Paul McCormick
printed and bound by McCurdy Printing & Typesetting Limited in Halifax, Nova Scotia, Canada

The front cover photo is a view of the Narrows from the Halifax side following the explosion. Courtesy of the City of Toronto Archives.

The back cover photo shows the Halifax Railroad Station after the explosion. Courtesy of the Public Archives of Nova Scotia.

Canadian Cataloguing in Publication Data

Sutow, Mary Pauline Murphy, 1954-

Worse than war

ISBN 0-920427-33-2

1. Halifax (N.S.) – History – Explosion, 1917 – Juvenile literature. * I. Title.

FC2346.4.S87 1992 j971.6'225 C92-098663-3
F1039.5.H17S87 1992

I wish to dedicate this book to the memory of those who died in and to those who survived the Halifax Explosion of 1917; and to my husband Elliott, and our three children Daniel, Eriel, and Patrick.

ACKNOWLEDGEMENTS

I wish to acknowledge the support and assistance I received from many people during this endeavour: Alan Ruffman of Geomarine Associates who has made it possible for me to use some of Arthur Lismer's drawings, who provided me with the material pertaining to the scientific aspects of the Explosion, and who reviewed my manuscript for accuracy; Mrs. Marjorie (Lismer) Bridges who granted me permission to use some of her father's works; Gerry Hickman, Geomarine Associates; David Flemming and Marvin Moore, Maritime Museum of the Atlantic; Margaret Campbell and staff at the Public Archives of Nova Scotia; Karen Teeple, City of Toronto Archives; Jan Anthony, Mount Saint Vincent University Art Gallery; Martin Carnaghan, United Memorial Church; City Historian, Lou Collins; Doug Frenette, St. Joseph's School; and finally, my husband Elliott for his moral and editorial support.

CHAPTER 1

HALIFAX AND WORLD WAR I

In December 1917, World War I was already more than 3 years old. Canada was fighting on the side of Great Britain, France, Russia and the United States, against the countries of Germany, Austria-Hungary, Bulgaria and Turkey. Over three years before, in August 1914, the Canadian Parliament had pledged its resources to the last person and the last dollar in support of the war. The rate at which people from Nova Scotia and its largest city, Halifax, volunteered to serve in the military was among the highest in Canada. Halifax would lose more than one thousand and three hundred men and women by the end of the war.

While it was mainly men who joined the navy and the army, women's role in the war effort in Canada was also important. Women carried on with life at home as normally as possible. They were told by one preacher that "This is women's war" and that the support they gave in many ways to the "brave sons of Canada" could not be measured.

Halifax was an important city in wartime. It had one of the best natural harbours in the world. For ships entering from the Atlantic Ocean, it had a wide outer harbour

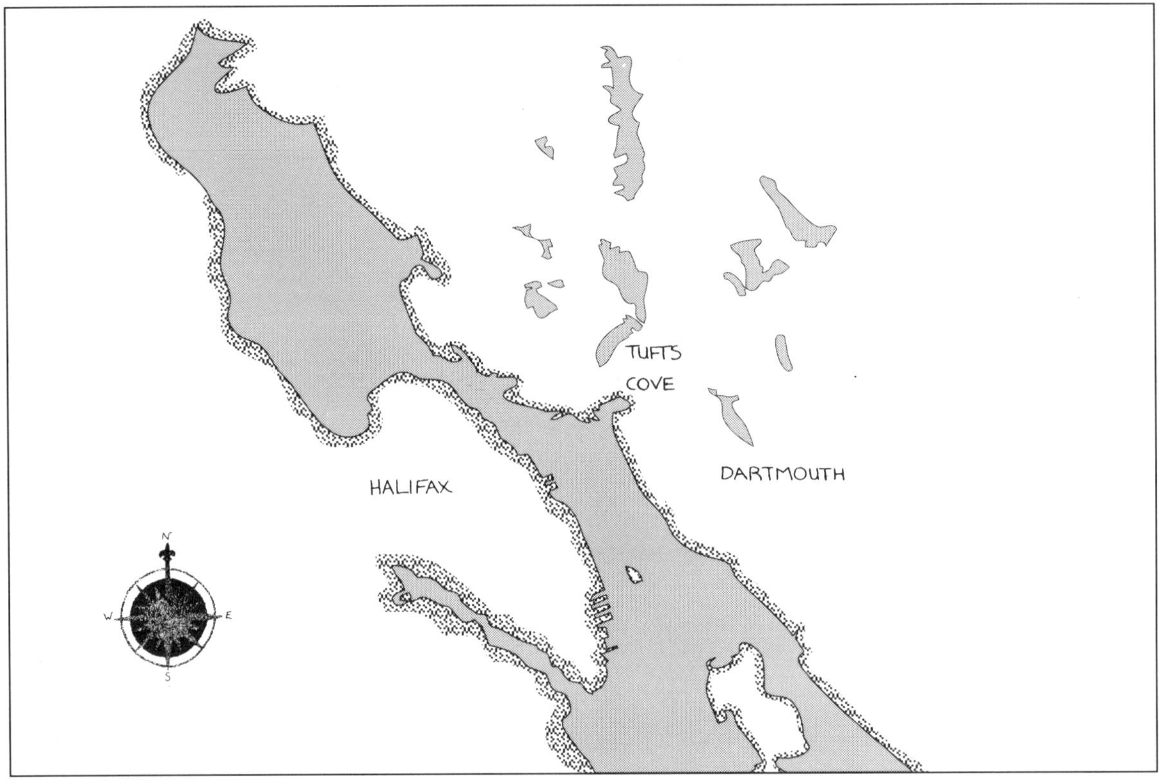

Halifax Harbour.

Map by Paul McCormick

and a deep channel called the Narrows. Once through the Narrows, a ship could then pass into a wide enclosed area of water, called the Bedford Basin. The Bedford Basin was large enough for a whole fleet of ships to gather. The harbour separated Halifax and Dartmouth, a smaller city, but one that had grown rapidly since 1900 and had several industries operating.

Because Halifax was such a great port, it served several important purposes during the war. Temporary shelters, such as wooden huts with tar-paper roofs, had to be built to house the many soldiers who gathered in the city while waiting to be sent to Europe. Most of the Canadian troops—it has been said three out of every four of the approximately half-million men and women—left for Europe through the port of Halifax.

Neutral ships bound for Europe were held in Halifax for inspection to ensure they were not contributing in some way to the military might of the enemy countries. Halifax served as a place to imprison enemy soldiers. One of the prisons used was the Citadel. Halifax also contributed its older and newly erected hospitals to treat wounded soldiers returning from Europe. One soldier was to say "Halifax is great—they treated us royally—we shall never forget it."

Before the war, Halifax was a busy port, handling two million tons of cargo a year. By 1917, it was handling seventeen million tons of cargo, much of it clothing, food, equipment and ammunition bound for the battle lines at the "Front."

Many soldiers and their families lived in the north end of the city in an area called Richmond. Thomas Raddall, in his book *Halifax, Warden of the North*, tells how Richmond got its name. Halifax, in its early days, got many shipments of flour from Richmond, Virginia, a city in the United States. The wharves where the flour was unloaded from the ships were called the Richmond Wharves. Eventually the surrounding area was given the name Richmond. The community of Richmond was built on and around the slopes of Fort Needham. On the waterfront below were the railroad, dockyards, piers and factories. Many people from Richmond worked in the waterfront area.

In 1917, the people of Halifax always had the war on their minds. They rose daily to work in their own ways to help their city do its part in the war. Families missed their loved ones who were fighting in Europe, and sadly, many found the names of friends and relatives who were killed, missing in action or seriously ill, listed in local newspapers.

On December 1, 1917, a French steamer ship, the *Mont Blanc*, left New York harbour bound for France. She was loaded with 2622 tons of explosives. She was to have been part of a convoy with British destroyer ships, but was left behind because she was too slow.

Mont Blanc, French Ammunition Ship

Maritime Museum of the Atlantic

The captain of the *Mont Blanc* had been told to go to Halifax with the dangerous cargo, to try and join another convoy that was planning to cross the Atlantic Ocean. The flying of a red flag was a rule for all ships on which explosives were being handled. Since the *Mont Blanc* was just carrying explosives and not handling them, the Captain did not fly the red flag. It was also suggested that not flying the red flag would avoid drawing the attention of the enemy to a prize target.

The *Mont Blanc* arrived at the mouth of Halifax harbour on December 4. She was boarded by a harbour pilot and inspected. She waited for permission to pass through the harbour into the Bedford Basin, where the convoy would form. The *Mont Blanc* was told she would have to wait until the next morning so that she could pass through in daylight. She dropped anchor in the outer harbour near McNabs Island.

On December 3, a Belgian Relief steamer, the *Imo*, was anchored in the Bedford Basin. During the afternoon of December 5 she was given permission by the harbour authorities to leave Halifax. A harbour pilot was sent on board to help. The *Imo* had to wait to have her coal bunkers filled. By that time it was almost dark. The *Imo* could not pass out of the harbour until the next day.

CHAPTER 2
THE COLLISION

Thursday morning, December 6, the *Imo* got under way at 8:10 without permission from harbour authorities. The captain was angry because of the delay the day before. The harbour pilot wanted to get her out to sea as soon as possible. The *Imo* gained speed quickly.

Rules of the sea require that ships in a channel or passageway should always pass "port to port" (left side to left side). This means that ships in a narrow channel should stay as far to the righthand side as possible. Once moving, ships in 1917 used whistle blasts to communicate with each other; now, of course, there is radio, but it was a new technology in 1917. To pass "port-to-port", the *Imo* should have been on the Halifax (west) side of the Narrows as she left the Basin and the *Mont Blanc* should have been on the Dartmouth (east) side of the channel.

As the *Imo* headed for the open sea, she met a United States freighter and later a tugboat, the *Stella Maris*. The position of these two vessels in the Narrows made it necessary for the *Imo* to move farther away from the Halifax shore than she should have been. In this position she was directly in the path of the *Mont*

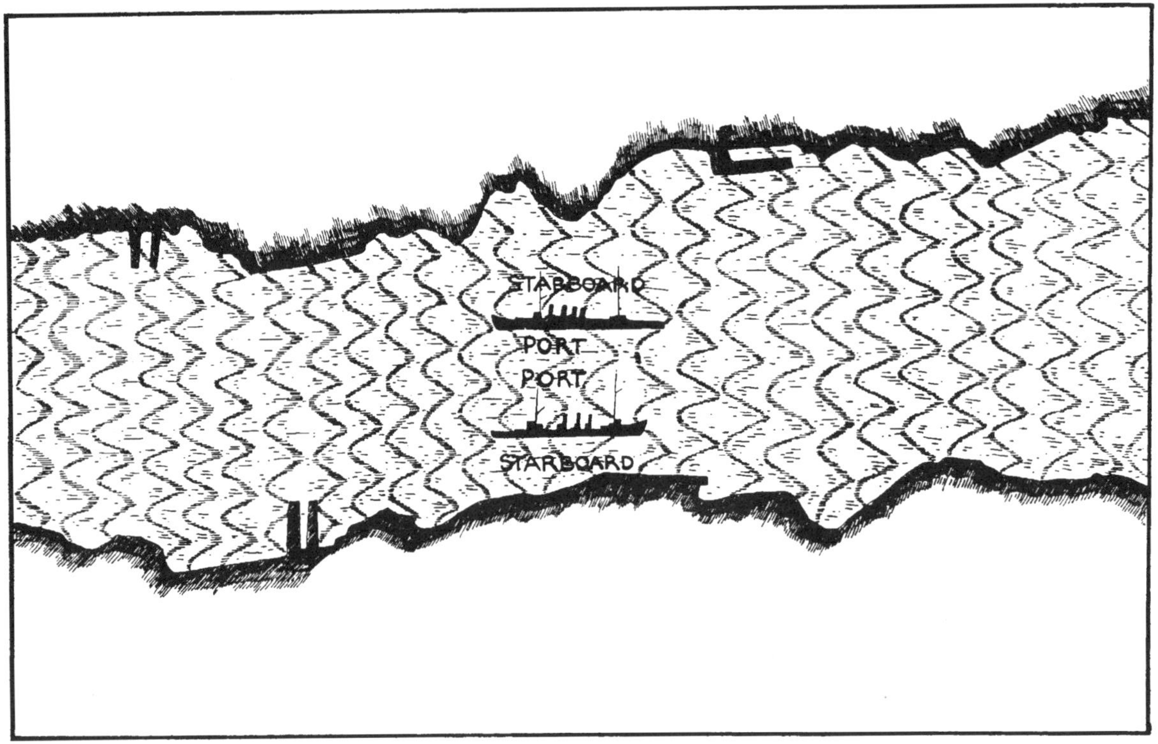

Rules of the Sea.

Marcus Adrian Jones

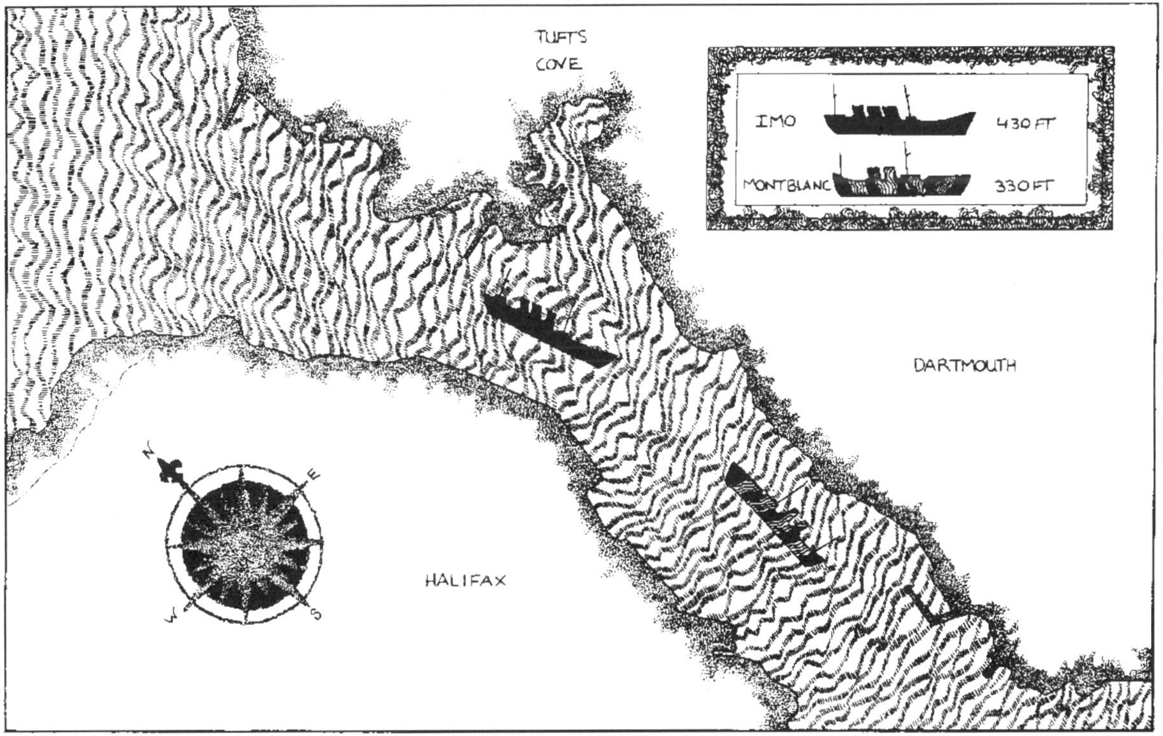

Positions and movement of the Mont Blanc and the Imo leading up to the collision.

Marcus Adrian Jones.

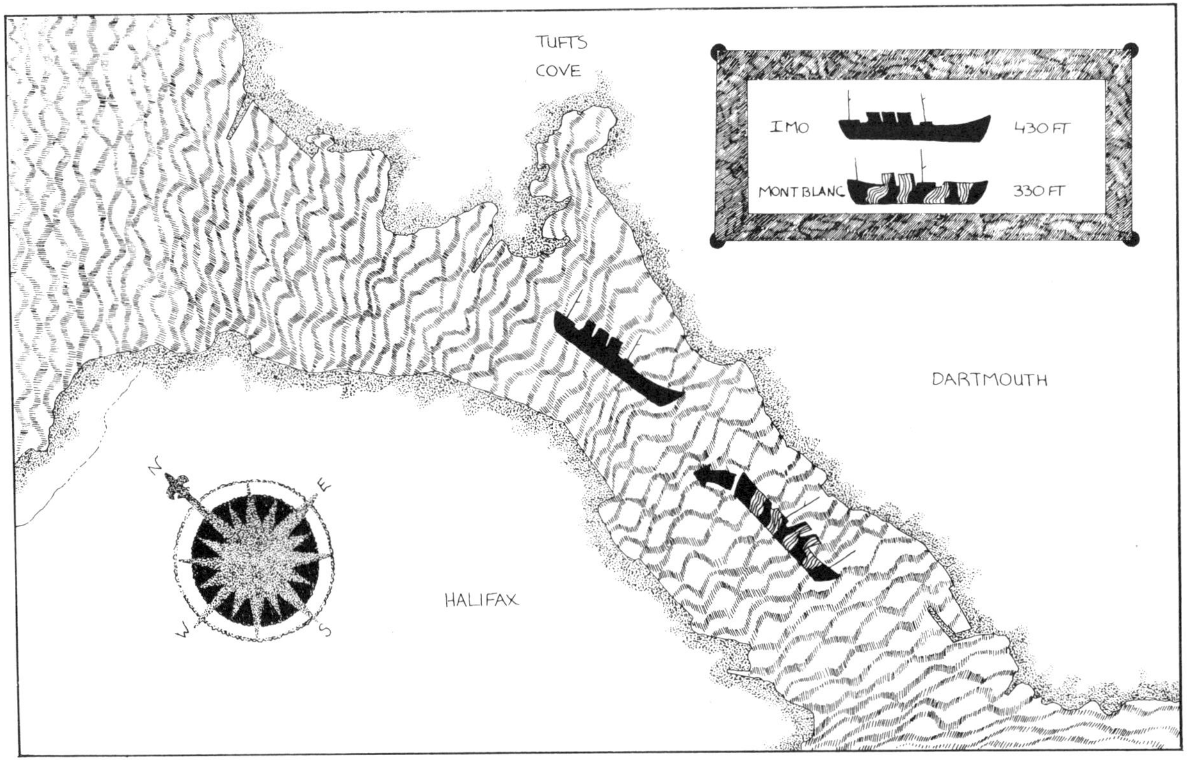
TUFTS
COVE
IMO
430 FT
MONT BLANC
330 FT
DARTMOUTH
N
E
W
S
HALIFAX

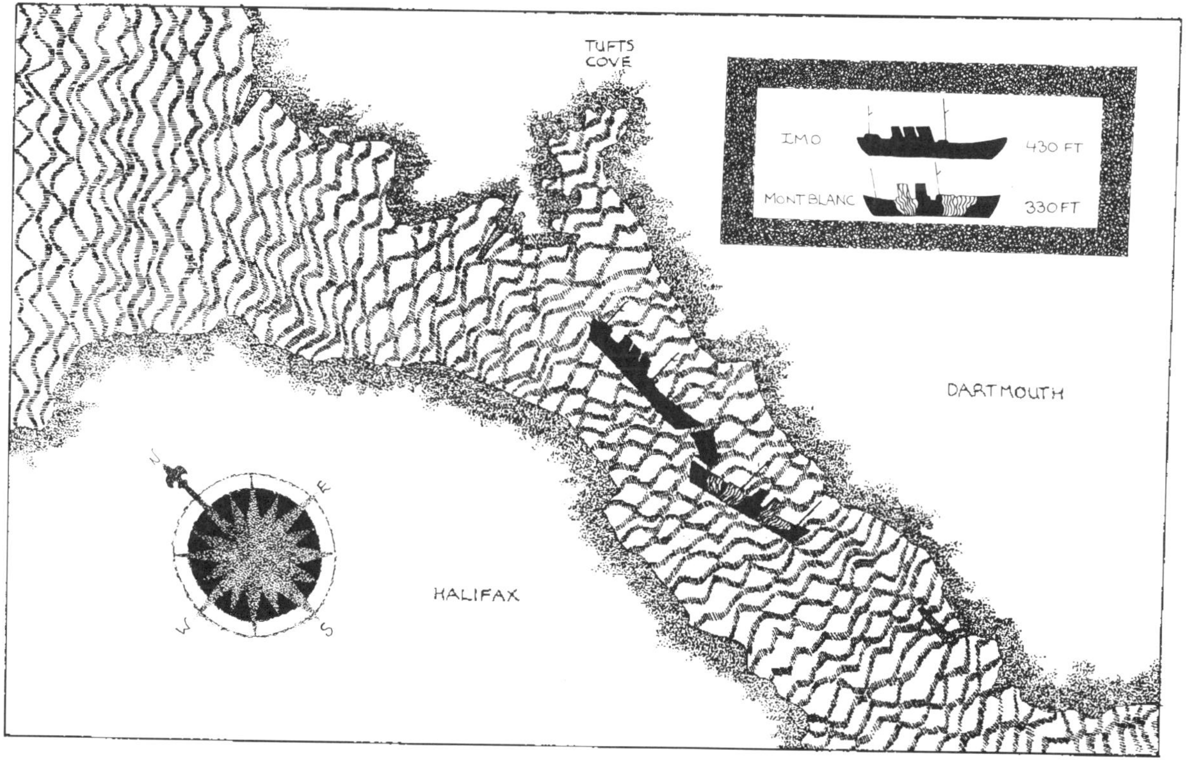
TUFTS COVE
IMO
430 FT
MONT BLANC
330 FT
DARTMOUTH
HALIFAX
N
E
W
S

Blanc, which had raised anchor at 7:30 a.m., after getting permission to proceed to the Bedford Basin.

The captain of the *Mont Blanc* gave one whistle blast to tell the *Imo* she was in the wrong place. The *Imo* blasted twice to say she was heading in the direction of Dartmouth, still across the path of the *Mont Blanc*. The *Mont Blanc* slowed and again gave one blast to say she was keeping to the Dartmouth side. The *Imo* again gave two blasts saying she intended to move toward the Dartmouth side.

Seeing trouble coming, the *Mont Blanc* swung to the left, which sent her back toward the middle of the Narrows (toward Halifax). This placed the ship parallel to the *Imo* for a short time. At this point, there was still enough time for the two ships to pass starboard to starboard.

The *Imo* blasted 3 times saying it was going to reverse its engines to slow and go astern. The *Mont Blanc* made ready to reverse as well. Because of the speed of the *Imo*, as she began to reverse her propellers, her bow (front) swung to the right. The *Imo* hit the *Mont Blanc's* bow at a place where some of the explosives were stored. After the *Imo* struck, the *Mont Blanc* stopped engines and lay damaged and drifting toward the Halifax shore.

Thick black smoke and then flames quickly rose from the bow of the *Mont Blanc* as she drifted slowly toward the Halifax piers. She was a time bomb with a short fuse. Her captain knew the ship would blow up and ordered her crew into lifeboats. Everyone escaped from the smoking ship and rowed to the Dartmouth shore, without ever giving the alarm that she was loaded with explosives. Once abandoned, the M*ont Blanc* eventually reached the Halifax shore and set fire to one of the piers.

CHAPTER 3

THE BLAST

Men, women and children all over Richmond and the Halifax waterfront stopped work and play to watch the unusual movements of the two ships in the harbour and then the burning *Mont Blanc*. People watched from windows in their Richmond homes, from factories, stores and streets.

Almost no one knew that the *Mont Blanc* was loaded with explosives and few understood the danger of a fire on a munitions ship. No one could warn the city of Halifax of the disaster about to happen. One man, Vincent Coleman, who worked in the train station and did know the danger, began to run from the area, but then went back to warn incoming trains by telegraph.

At four minutes and thirty-five seconds past nine o'clock on December 6, 1917, the *Mont Blanc* exploded with a mighty force.

A Halifax scientist, Alan Ruffman, has described the Explosion as a scientific event. An explosion is a chemical fire that burns rapidly through explosives that serve as the fuel. It took only one-fiftieth of a second for the fire to use all the fuel on the *Mont Blanc* and to convert it to gas and heat energy. Only a small amount of this energy was needed to shatter the *Mont Blanc* and spew its pieces in all directions. A piece of her anchor was later found more than two miles from the site of the explosion. The cannon on its stern went the other way and landed in Dartmouth. Immediately, a large black cloud rose into the clear morning air, casting darkness all around and raining debris on all below.

Vincent Coleman, telegraph operator.
Public Archives of Nova Scotia.

The energy produced resulted in a ground or seismic wave, a water wave and an air wave. The ground wave travelled the farthest and its vibrations were felt by people in other parts of Nova Scotia and in New Brunswick and Prince Edward Island as far away as 450 kilometers.

The water wave travelled the shortest distance. The Explosion created a hole in the harbour floor and sent water and boulders into the air. The effect of the

**Thirteen miles from the Halifax Explosion—
How it appeared at that distance.**
U.S. National Archives/Reproduced by Julian Beveridge

water being blown outwards and into the air and then rushing back into the hole caused a short-lived 'tidal wave' of water to rush up onto the shorelines. Ships and boats in the area were tossed about like bobbers. One man in a rowboat found himself sitting high and dry on the Dartmouth shore. The water wave lost its energy quite quickly and did little damage compared with the air wave.

It was the air wave that drove down all in its path. Houses, stores, factories, boat yards and piers were broken and flattened like toys under a big boot. Homes fell in on the household stoves inside, causing fires. Very soon the flames made it difficult to save people in the crushed buildings.

In Halifax, more than nineteen hundred people were killed at once and more

Wrecked Home in Richmond

Public Archives of Nova Scotia.

Help the Halifax Blind, Reproduction of a pen and ink drawing by Arthur Lismer.
Canadian Courier Vol. XXIII, No. 10, February 16, 1918a/A. Ruffman, Geomarine Associates Ltd.

than nine thousand people were injured. Many injured were blinded or had face and eye injuries. This was because pieces of window glass flew into their faces as they had watched the burning ship with great interest.

In Dartmouth, forty people were killed immediately, with about as many more dying in the ten days that followed, in Explosion related deaths. One hundred and sixty-two houses were left unfit to live in and an entire Micmac Indian encampment on the Dartmouth shore was wiped out.

More than thirty thousand people in Halifax were left with either no home or one unfit to live in. Those able to travel were taken by train to towns and communities elsewhere in Nova Scotia until their homes could be repaired or rebuilt.

The railroad yards and trains in Richmond were ruined. The roof of the main station at North Street had caved in with many lives lost. Vincent Coleman, who had stayed behind to warn incoming trains, died right away.

Alice Bardsley has written of one girl, Barbara Orr, who was thrown up the side of Fort Needham, yet still survived. Barbara Orr was lucky. Many others who had been blown uphill were killed. The girl's family all died from the force of the Explosion.

Many people just disappeared, never to be found again.

A soldier who had returned from the front in France was to say that what he saw in Halifax after the Explosion was worse than anything he had seen on the battlefield.

Homeless, reproduction of pen and ink drawing by Arthur Lismer.
The Drama of a City, published by Gerald E. Weir, New York City 1918c/A. Ruffman, Geomarine Associates Ltd.

Arthur Lismer, one of the famous Group of Seven artists, was working in Halifax

Halifax Railroad Station, in which 60 persons were killed by falling roof.

Public Archives of Nova Scotia

at the time. He did some hurried sketches for a Toronto magazine. One of these sketches had a caption that read in part ". . . not somewhere in France but almost anywhere in Halifax." More Haligonians were killed in thc Explosion than were killed in World War I.

CHAPTER 4

THE RESCUE

The people who were able began immediately to help those in need. Soldiers and sailors and people from other parts of Halifax and Dartmouth started to do all they could to help the injured. The existing hospitals filled up very quickly as many wounded and moaning men, women, and children crawled, walked or were carried out of the north ends of Halifax and Dartmouth. Extra hospitals were made out of other buildings and houses. Theatres, movie houses, schools and churches became relief centres. Doctors, nurses and all those able began the great task of attending to the thousands who needed medical care.

A school in Halifax was set up to serve as a mortuary, where the dead people were taken. By December 9th, local newspapers reported that two hundred and ninety-nine dead were taken to the temporary morgue. By December 13th, the number rose to eight hundred. Of these, four hundred and fifty had been identified by family and friends. One hundred and twenty-five could not even be identified. Nearly one hundred of these unidentified were buried as a group on December 17.

The Hour of Horror in Devastated Richmond, Reproduction of a pen and ink drawing by Arthur Lismer.

The Drama of a City, published by Gerald E. Weir, New York City 1918b/A. Ruffman, Geomarine Associates Ltd.

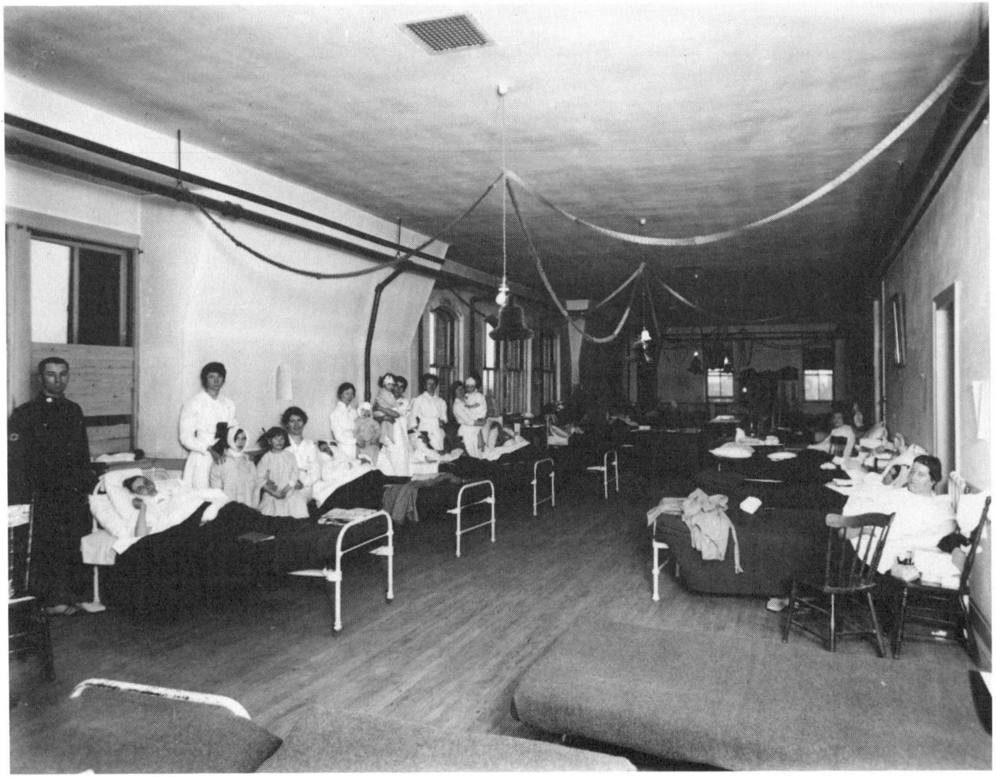

Temporary Hospital

James Collection, City of Toronto Archives.

Unidentified dead

James Collection, City of Toronto Archives.

Help from other parts of Nova Scotia and Canada and from the United States came as soon as trains and ships could get to the area. Badly needed food, furniture, clothing, bedding and medical supplies were sent. The *Halifax Chronicle* on December 7th gave news of a special relief train from New York City. It carried six hundred men, $15,000 worth of tools, $150,000 worth of lumber, one thousand portable houses, and twenty-five trucks. As well, a special team of salvage workers was to be directed in the restoration and construction work by twenty expert engineers.

Depots for the distribution of food, clothing, furniture and other necessities, were set up in many places in the city. Residents and businesses in parts of Halifax not as severely affected as the North End contributed their time, homes, food and automobiles as the relief work proceeded. Local bakeries sent out bread teams that were equipped with bells. Once residents of the affected areas heard the bells, they were to come to their doors and receive some bread for themselves and for whoever was left of family and friends.

Doctors, nurses and other health workers were sent in to help the medical teams in the area. They were badly needed. Still, many people died while waiting to be treated.

Help in the form of money came from the governments of Canada, and other

Furniture Depot

Public Archives of Nova Scotia.

In the blizzard of the second day, Reproduction of a pen and ink drawing by Arthur Lismer

The Drama of a City, published by Gerald E. Weir, New York City 1918d/A. Ruffman, Geomarine Associates Ltd.

countries such as Australia, Britain, South America, China, the United States, Newfoundland (not yet a Canadian province), and New Zealand. The French company that had owned the *Mont Blanc* sent $10,000.

The weather in Halifax made the rescue effort even harder. The day after the Explosion a major snowstorm left sixteen inches of snow on the ruined city. On Sunday, December 9, the snow melted due to spring-like temperatures, only to have the mess of water and slush freeze again on the next day. It was as though Halifax was being battered from all sides. The war in Europe was not important during this time. It was far from most people's minds as they fought to put their lives in some sort of order.

CHAPTER 5

CHRISTMAS 1917

A local newspaper that was on the newstand before the explosion that morning of December 6 carried an article debating whether Christmas should be celebrated as usual. Due to the drains on the people and the country it could be said that the extras and treats which go with Christmas were too extravagant in this time of war. The writer of the piece, however, encouraged people to "give children their Christmas." If the war continued, then perhaps trimmings in life and at Christmas, in particular, would have to be cut out later. But so far as children were concerned, in 1917 Christmas should be little different from other years.

Little did the writer of that piece know that on that very day events would occur that would change Christmas, not only in 1917, but for all future years for those who survived the Explosion.

After the Explosion, relief workers vowed to do all they could to bring some kind of Christmas celebration to the children who were suffering from injury, from being homeless and from being without parents and family members.

On the evening of December 21, Santa Claus Limited was formed. The

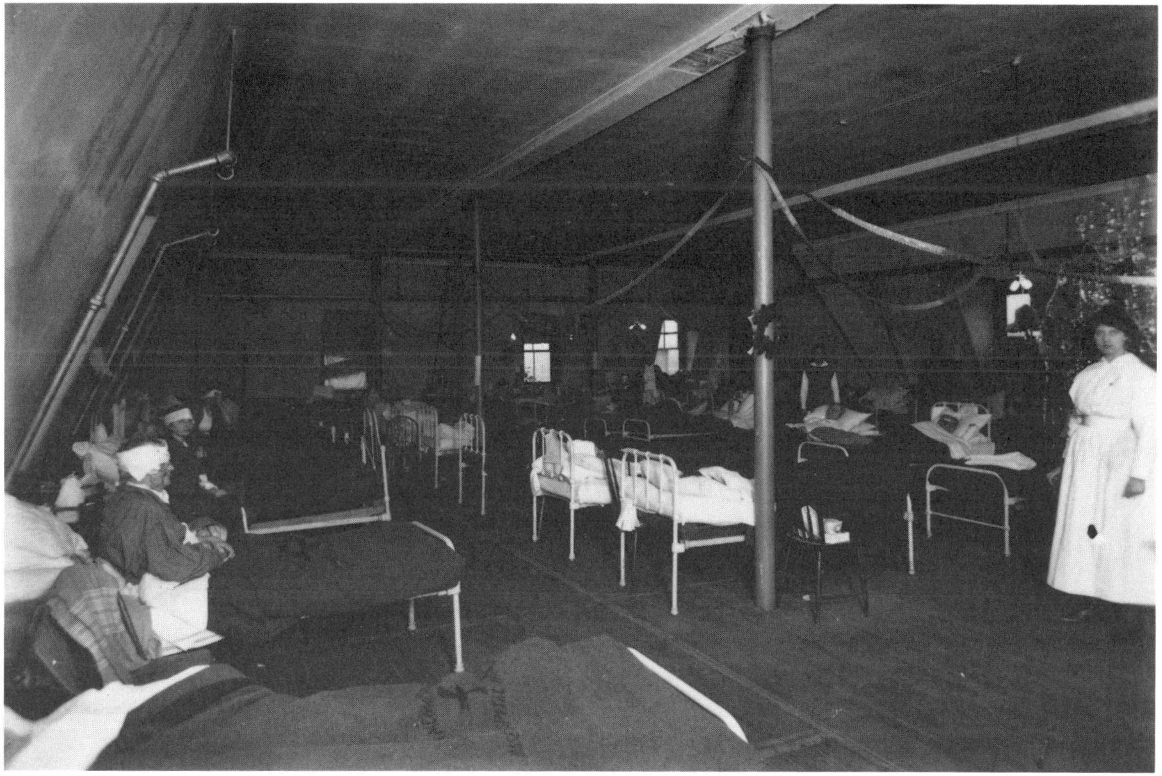

Christmas in Hospital

James Collection, City of Toronto Archives

organization appealed for help in the form of money, automobiles and workers to help in its task. It appealed especially to those persons who had not yet fully contributed to the relief efforts.

Santa Claus Limited succeeded in its work. On December 26, a newspaper read "and so it was everywhere a day of joy," at least as far as possible. Gifts, fruitcake, and candy were distributed to about ten thousand children. Those able gave gifts of their help to those in need.

One woman wrote that "no one had much desire for giving or receiving outside of relief work." It was that generous giving in many forms that made up the spirit of Christmas in Halifax that year. For the givers, it was being able in some way to cast some light into the darkness surrounding so many people. For the receivers, it was to be remembered forever as a drop of joy in the pool of sadness, pain and misery.

CHAPTER 6

PLACING THE BLAME

Besides the great pain and loss felt by the people of Halifax, there was a feeling of anger. They felt anger toward the person or persons who had done this to them and their city. Right after the explosion many thought the Germans were at fault. Some people still believe it. German people living in the city were subjected to racist beatings and abuse. The French crew of the *Mont Blanc* was also abused by a population looking for a scapegoat. The citizens needed someone to blame for the devastation that lay all around them.

On December 12, six days after the Explosion, an inquiry into the collision began in a broken-down courtroom. A Nova Scotia Court judge began to hear and examine the facts. Among more than fifty people who told their stories were the captain and pilot of the *Mont Blanc*. The captain and pilot of the *Imo* had died because of the Explosion.

On February 4, 1918, the judge made his report. He placed all the blame on the captain and pilot of the *Mont Blanc*. They were both arrested. Many agreed

The Imo

Public Archives of Nova Scotia.

with the judge. Others did not.

In May 1918, five judges of a higher court, the Supreme Court of Canada, gave their decision. Two of the five found the *Mont Blanc* to blame, two found the *Imo* to blame. One judge stated his belief that both ships were equally to blame.

An English court was asked to review the facts. The decision of this court was that the *Imo* and the *Mont Blanc* were equally to blame. The captain and the pilot of the *Mont Blanc* were allowed to do their jobs again. The *Imo* was repaired, renamed and returned to service. She sank in December 1921, after hitting a reef in the South Atlantic Ocean, near the Falkland Islands.

More than the captains and pilots on the ships were found to be guilty in this, the largest ever accidental explosion in the world. The Halifax pilotage commission became the subject of an inquiry. It was revealed that the rules governing the movement of ships in the harbour were those that had been applied two hundred years earlier when sailing vessels used the harbour. These rules were certainly no longer suitable to properly handle ship traffic in 1917, especially because of the war.

In the twenty-five years before the Explosion, at least two Pilotage Commissioners had urged the Canadian Department of Marine and Fisheries to examine and change the outdated regulations. Their requests went unanswered and the Halifax

Explosion may have resulted partly from this neglect. Even as the inquiry proceeded, on February 17, 1918, a large ship with thousands of passengers only narrowly escaped catastrophe because of the poor management of traffic in the harbour. On March 7, 1918, a steamship loaded with explosives was left anchored in the fairway of the harbour for seven hours during a snowstorm. Incoming ships unable to see the ship could have struck her and another catastrophe might have occurred.

The inquiry being conducted was badly needed. Other vital considerations in the inquiry were the training of pilots, the number of hours per day pilots were allowed to work and whether their eyesight had been properly examined.

In time, the rules for the use of the harbour were updated, making Halifax a much safer harbour. By then, however, it was too late for the thousands who were killed, injured or orphaned by the Explosion.

CHAPTER 7

REMEMBERING THE PAST

The severely damaged community of Richmond struggled to go on as well as possible. The road network was redesigned and a new plan was made. Houses, schools, stores and factories were rebuilt or repaired. The money sent to Halifax after the Explosion was used for many years and is still being used through the Department of Veterans Affairs to help and support the injured.

Two of the four churches that had been destroyed, the Kaye Street Methodist Church and Grove Presbyterian Church, joined together. The first service after the union was held in a tarpaper structure on March 17, 1918. A new building to house the United Memorial Church was finished in 1921, complete with bell tower and a carillon of ten bronze bells. This United Church, at the end of the new Union Street, was one of the first "United Churches" in Canada. It predated the union of the Presbyterian and Methodist Churches in Canada by eight to nine years. Barbara Orr, the girl who had been blown up the side of Fort Needham, presented the bells to the church in memory of her parents and her five brothers and sisters who had died in the Explosion. She was the first to play the bells. Other

Children getting food from a relief station

Public Archives of Nova Scotia.

Monument to those who perished in the Halifax Explosion, Fort Needham Memorial Park
Julian Beveridge Photographic Services

parts of the church were presented by survivors in memory of lost or killed loved ones.

A long overdue monument to the Halifax Explosion was finally built in 1985 on top of Fort Needham, just above the site where, sixty-eight years before, the burning *Mont Blanc* lay waiting to erupt in fury. The monument became the new home for the bronze bells that had first hung in the bell tower of the United Memorial Church. A dedication ceremony was held June 1985.

A service is now held on December 6 of each year in memory of those who died from, or whose lives were changed forever by the great Halifax Explosion of 1917.

SOURCES AND REFERENCES

Alice Bardsley "Memories From the Bells". *Atlantic Advocate*. 75:41-43. December, 1984.

Michael J. Bird *The Town That Died*. London: Souvenir Press, 1962.

Paul A. Erickson *Halifax's Other Hill: Fort Needham From Earliest Times*. Halifax: St. Mary's University, 1984.

Charles Bruce Ferguson "The Halifax Explosion". *Journal of Education* 17:25-31. February, 1968.

R.L. Fraser "My Experience of the Halifax Explosion". *Atlantic Advocate* 57:49-52. January, 1967.

Webb Garrison *Disasters That Made History*. New York: Abington Press, 1973.

Janet F. Kitz *Shattered City: The Halifax Explosion and the Road to Recovery*. Halifax: Nimbus Publishing Ltd., 1989.

Hugh MacLennon *Barometer Rising*. New York: Sloan and Pearce, 1949.

John Patrick Martin *Story of Dartmouth*. Halifax: privately printed for author, 1957.

Graham Metson *The Halifax Explosion of December 6, 1917*. Toronto: McGraw-Hill Ryerson Ltd., 1978.

Thomas H. Raddall *Halifax, Warden of the North*. Toronto: McClelland and Stewart Ltd., 1971.

Alan Ruffman *The Halifax Explosion: Realities and Myths*. The working text for three talks given on CBC Radio's Information Morning Show in Halifax on the occasion of the 70th anniversary of the Halifax Explosion of December 6, 1917. CBC Radio Station CBHT, Halifax, Nova Scotia, December 2, 3, 4, 1987, at 0740 AST, unpublished manuscript, 29 pp., 1987.

Alan Ruffman *Preliminary Annotated Research Bibliography of Published Material on Arthur Lismer, His Time in Halifax, and on the December 6, 1917 Halifax Explosion*. Geomarine Associates Ltd., Halifax, Nova Scotia, Project 83-29 (Lismer) unpublished Manuscript, 22 pp., 1990.

Alan Ruffman *Grim Visions, Arthur Lismer and the Halifax Explosion*. Art Gallery, Mount Saint Vincent University, Halifax, Nova Scotia, Design Jan Anthony/James Graham, 4 pp., 1990.

Halifax Explosion Papers. Public Archives of Nova Scotia.

ANNOTATED REFERENCES FOR SELECTED ILLUSTRATIONS

p. 16 Explosion Cloud

Anonymous, 1917, Thirteen Miles from the Halifax Explosion—How it appeared at that distance. U.S. National Archives, National Archives and Records Administration, Washington, D.C., U.S.A. 9.5 x 7.5 in. approximately, print no. 165-WW-158-A15.

p. 18 Arthur Lismer. 1918a. Help the Halifax Blind. Canadian Courier, Vol. XXIII, No. 10, February 16, p. 11, cols. 1.8-3.2, pen and ink drawing, 9.7 x 11.3 cm. as printed.

p. 19 Arthur Lismer. 1918c. Homeless. The Drama of a City, The Story of Stricken Halifax. Published by Gerald E. Weir, New York City, printed at 17-27 Vandewater Street, New York City, U.S.A., facing p. 20, pen and ink drawing, 15.3 x 11.3 cm. as printed.

p. 23 Arthur Lismer, 1918b. The Hour of Horror in Devastated Richmond. The Drama of a City, The Story of Stricken Halifax. Published by Gerald E. Weir, New York City, printed at 17-27 Vandewater Street, New York City, New York, U.S.A., facing p. 14, pen and ink drawing, 12.0 x 20.0 cm. as printed, with the top of the drawing to the spine and facing to the right.

p. 28 Arthur Lismer. 1918d. In the blizzard of the second day (December 7, 1917) many scenes of horror were covered from view, but the work of searching the ruins went on and the mutilated bodies encrusted with snow were piled on sleds to be drawn to the morgues. The Drama of a City, The Story of Stricken Halifax. Published by Gerald E. Weir, New York City, printed at 17-27 Vandewater Street, New York City, New York, U.S.A., facing p. 48, pan and ink drawing, 8.1 x 10.5 cm. as printed.